INTRODUCTION

Classical elegance has been an attribute of several combat aircraft throughout aviation history, but aesthetics have not always been complemented by the more martial virtues that make such machines truly outstanding. The Hawker Hunter is an exception. To many people, the early variants, unencumbered by 'dog-tooth' wing leading edges and vast gun blisters, represented aerodynamic perfection, but even when military realities forced a number of external modifications on to the airframe the Hunter's graceful lines were never disguised. Nearly 2000 aircraft were produced, almost half of these specifically for service with overseas air forces, and as the Hunter was gradually relinquished in favour of more modern types during the 1960s the surplus stock was snapped up by foreign customers as fast as the machines could be refurbished. The RAF and the Royal Navy still operate a few two-seaters, and it is likely that some overseas Hunters will still be in front-line service ten years hence. Nevertheless, the aircraft's early days were not trouble-free . . .

TRIALS AND TRIBULATIONS

The first Hunter prototype, designate[illegible] P.1067 by Hawker, began to tak[illegible] shape in early 1950, the design havin[illegible] evolved through a series of studies and experimental aircraft originating with the P.1040 Sea Hawk. In 1947, under the leadership of the legendary Sir Sydney Camm, Hawker's design staff had put forward proposals for a high-speed, single-seat interceptor built around the brand new Rolls-Royce Avon turbojet, and the following year Specification F.3/48 had been issued authorising Camm to proceed. The final requirements involved the construction of three trials aircraft; the Avon and the Armstrong Siddeley Sapphire as alternative powerplants; a maximum speed of Mach 0.94; and a battery of four of the new 30mm Aden cannon. By the time the P.1067 was ready for its maiden flight, Hawker had received a production contract for the aircraft.

The second and third prototypes were flying by the end of 1952, the latter with the Sapphire engine, and in the meantime the original P.1067 had been modified for, and had success-

[illegible] air [illegible]r, a [illegible]ning

First, the same [illegible]amic splendour that gave the Hunter its efficiency in the air brought with it a major drawback – inadequate deceleration. This question was resolved, via a number of experimental installations, by fitting a single large ventral air brake, complemented in later years by a tail-mounted parachute to assist deceleration on the ground. Second, difficulties with pitch control eventually led to the addition of leading-edge extensions to the outboard wing panels. Third, and more seriously, the ejection of spent ammunition after the Aden cannon had been fired

Below: A superb study of the P.1067 prototype. The first flight took place from A&AEE Boscombe Down on 20 July 1951 and the aircraft, in the hands of Hawker's chief test pilot Neville Duke, remained aloft for some three-quarters of an hour. WB188 here carries an overall glossy pale green paint finish. *British Aerospace*

threatened, and on a number of occasions actually caused, damage to the airframe, and the problem was solved only by collecting the spent cartridge links in a pair of external containers mounted behind the guns. Fourth, the aircraft proved to be critically short on range, owing to insufficient fuel capacity – a shortcoming which would also afflict the Hunter's successor, the BAC Lightning (see Aeroguide 8) – and a satisfactory answer could come only by adopting underwing fuel tanks, which in turn would require some measure of wing redesign and which therefore could not be effected in early Hunter variants.

Thus was the aerodynamic purity of the aircraft steadily diminished. A further difficulty was engine flame-out, brought about by gas generated by firing the cannon causing turbulence around the main intakes, resulting in a phenomenon known as compressor surge. Considering the ejection problems already referred to and the fact that pitch control was further adversely affected when the weapons were operated, the Hunter and the Aden cannon were, at first, not exactly a dream marriage.

THE DEFINITIVE HUNTER

The difficulties were gradually overcome, and as the years passed the Hunter proved itself to be a highly efficient and extremely popular aeroplane. The Avon-powered Mk 1, and its Sapphire-engined equivalent, the Mk 2, entered service in 1954 and 1955 respectively and were succeeded each in turn by the Mks 4 and 5; the Mk 3 designation was given to the record-breaking first prototype. The Mks 4 and 5 differed principally in having uprated engines, a limited additional fuel capacity by way of integral wing tanks forward of the main wheel bays, and provision for two wing pylons; further modifications during the production run allowed F.4s two extra pylons (or, optionally, rocket racks).

The introduction of a 10,000lb thrust Avon in place of the 7500lb 100-Series engine of the Mk 4 resulted in what is generally considered to be the definitive Hunter, the F Mk 6, which as well as equipping the RAF also formed the basis for the major export variant. Many F.6s were subsequently modified for the ground-attack role as Mk 9s, the main changes being a provision for 230gal tanks on the inboard wing stations, the recontouring of the rear fuselage to incorporate a parachute brake, and various internal improvements chiefly concerned with air-conditioning, oxygen and avionics.

A second line of development brought about the production of two-seat Hunters. Only one principal trainer version was built, the T Mk 7,

Above: During the first half of 1952 WB188 was modified to accept an afterburning Avon powerplant, and on 7 September that year the aircraft, now sporting overall bright red paintwork, was flown by Neville Duke at 727.6mph, thereby establishing a new world absolute air speed record. This photograph shows the reshaped tailpipe, the new side-mounted air brakes (note the hinge fairing immediately beneath the serial number) and the recontoured nose. *British Aerospace*

Left: The elegant lines of the Hunter are displayed to advantage in this photograph of the second F Mk 6. This particular machine, in common with several other early Mk 6s, was in effect a re-engined Mk 1 airframe (which no doubt explains the contrasting finish of the rear fuselage) and was retained by the manufacturers for trials purposes. The aircraft lacks the extended wing leading edges which would become characteristic of the mark. *British Aerospace*

although some of these were equipped with arrester hooks for Royal Navy use as T.8s. The T.7 was essentially an F.4 with a redesigned forward fuselage to accommodate instructor and pilot side by side, but the gun armament was reduced to a single weapon on the starboard side and a parachute brake was added at the tail; many of these aircraft were in fact converted from redundant F.4 airframes.

The Mk 10 was a reconnaissance Hunter based on the F.6 and equipped with three cameras in the nose, plus the Mk 9 improvements described above. The Mk 11, for Royal Navy use, was an unarmed single-seat trainer equipped with an arrester hook and, in the case of a few aircraft, with cameras; all were converted from F.4s. Finally, the sole Mk 12 was an F.6-derived two-seater intended as a prototype for a training aircraft that would be used in conjunction with the TSR.2 bomber programme.

REFURBISHING AND RESALE

The Hunter has been exported on a scale as yet unmatched by any other postwar British military aircraft. US-funded licence-production of Mk 4s and Mk 6s took place in the Low Countries, and Mk 4 airframes formed the basis of sales to Denmark, Peru and Sweden. Hunters with 200-Series Avons were purchased in large numbers by India and Switzerland, and on a more modest scale by Iraq, Jordan and the Lebanon. Following the withdrawal from front-line service of RAF, Dutch and Belgian Hunters in the early 1960s, a continuing programme of refurbishing and resale was undertaken by the parent company, new recipients of the aircraft being Abu Dhabi, Chile, Kenya, Kuwait, Oman, Qatar, Rhodesia (Zimbabwe), Saudi Arabia and Singapore.

After thirty years of service, the single-seat Hunter was officially retired from the Royal Air Force in the summer of 1984, and only a handful of T.7s remain in RAF colours, most serving with the Buccaneer OCU at Lossiemouth. Still with the Royal Navy are the GA.11 and T.8 trainers, some of the latter further modified for such purposes as TACAN and Blue Fox radar (Sea Harrier) training.

The lasting memory for those fortunate enough to have witnessed such events, however, will doubtless be the sight of large numbers of Hunters performing precision aerobatics – No 111 Squadron's *Black Arrows* and their No 92 Squadron successors the *Blue Diamonds* in the late 1950s and early 1960s. No aircraft has been more suited to such displays – nor looked better engaged in them!

Above: A production Hunter F Mk 6 poses for the camera, showing the large 230gal drop tank and 3in rocket battery carried underwing. Although here wearing British insignia, XK161 in fact never saw RAF service: it was one of the many Hunters cancelled by the UK government under the now-infamous 1957 Defence White Paper, and the aircraft quickly found its way into the Indian Air Force, redesignated Mk 56. *British Aerospace*

Right: A formation of Hunters from No 4 Flying Training School, photographed in 1975. The aircraft furthest from the camera is a T Mk 7, the remainder being F Mk 6s; all wear the standard red/white/Light Aircraft Grey trainer scheme of the day, and all are equipped with the smaller, 100gal underwing drop tanks. No 4 FTS flew its Hunters alongside the School's Folland Gnat advanced trainers at RAF Valley until the arrival of the new British Aerospace Hawk (see AEROGUIDE 1) in the late 1970s. *Ministry of Defence (RAF)*

Top: Another of the early production Hunter F.6s employed by the manufacturers as a test aircraft was WW594; along with XF378 (see photograph on page 33), it was assigned to a programme investigating the possibility of equipping the Hunter with De Havilland Firestreak infra-red air-to-air guided missiles. Modifications included the installation of Mk 20 radar equipment, which in turn required a redesigned nose and the deletion of half the cannon armament. *British Aerospace*
Above: The Hunter has proven to be the outstanding export success among postwar British military aircraft, equipping nineteen overseas air forces. An early customer was India, which eventually received about 240 single- and two-seaters. Part of the initial batch of Mk 56s is shown. *British Aerospace*
Below: The Royal Netherlands Air Force also introduced the Hunter into service in large numbers; most of the aircraft were licence-built, but the first batch, T Mk 7s, was manufactured by Hawker. Evident in this photograph are the two-seater's new upper-fuselage contours. *British Aerospace*

Below: The Royal Air Force's single-seat Hunters were officially retired in July 1984, the last users being No 1 Tactical Weapons Unit at RAF Brawdy. XE606, shown here, was officially designated an F.6A, but, with its parachute brake fairing above the tailpipe and the jettison gun fairings over the outboard wing pylon positions, it was externally almost indistinguishable from the ground-attack FGA Mk 9. The Mk 6As were Mk 6s modified in service to carry 230gal tanks to give them range compatability with the FGA.9s, but they lacked the latter's full weapons capability and their 'tropical' modifications (such as the special air-conditioning).

Bottom: Hunter T Mk 7 XL617, still in service with No 1 TWU during the autumn of 1984. The broad two-place canopy and its support strut are very evident.

Opposite page top: A view of the port forward fuselage of XE606 highlights the Hunter's exceptionally clean aero-dynamic lines.
Opposite page bottom: A close-up of the port side of the nose, showing the positions of the 30mm Aden cannon troughs and the simple but functional nature of the pilot's access steps. Atop the nose, forward, is the aperture for the aircraft's camera gun. Note that in several places the paint-work of adjacent panels does not match up, suggesting the use of components from more than one airframe.
Above: The nose from the starboard side. The large repainted panel beneath the windscreen contrasts with the 'blued' Dark Green elsewhere.
Below: Detail view of the centre fuselage, port side, showing the Hunter's characteristic rear transport joint (right), the wing root trailing edge fillet and the various engine louvres.

Right and below: Two views showing the general arrangement of the rear fuselage and tail surfaces. The prominent ventral feature is the large air brake, the development of which was the subject of intensive trials in the Hunter's early days. Also of interest in these photographs are the unsophisticated but very effective rudder clamp and the fact that the tailpipe seemingly originates from another airframe. The aircraft in the background are No 1 TWU BAe Hawks.
Opposite page top: The Hunter's fin-mounted tailplane was also the subject of considerable research by Hawker: the original P.1067 mock-up showed it fitted to the fin top. The fairing at the trailing edge root was incorporated into the design after prototype testing had revealed some problems of vibration.
Opposite page bottom: Although assigned to No 1 TWU, many of the Brawdy Hunters displayed the red and white markings of their 'shadow' squadron, No 79.

XE606

Opposite page top: The tail braking parachute is a feature of Hunter two-seat trainers, as it was of those Mk 6s either converted to FGA.9s or brought up to Mk 6A standards and of the small batch of Mk 6s converted to FR Mk 10s for the reconnaissance role. The 'chute is housed above the tailpipe and enclosed by a pair of doors, as shown in this view.
Opposite page bottom: A close-up of the fin trailing edge 'bullet' fairing, showing the navigation light at its tip.
Above: The subtle contours of the parachute fairing, the characteristic 'hood' formed over the tailpipe well in evidence.
Below: Lower rear fuselage details, including the tail bumper and vent pipe (the latter fitted to the port side only); the air flow pattern is very obvious! The stencilled lettering is black, and note the rear fuselage fastener behind the vent.

Above: A rare glimpse inside the ventral air brake; the brake is normally retracted flush with the fuselage when the aircraft is on the ground. Main features visible here are the actuating jack and the interior reinforcing channels.
Right: The F.6's engine starter access door, situated beneath the fuselage, between the main gear bays. The T.7's door differs, having three Coffman cartridges mounted on its inside surface; these were not required for the F.6's 200-Series Avon liquid-fuel starter.
Below: Starboard intake fairing, showing the fixed air bleed outlet and, nearer the leading edge, the spring-loaded auxiliary intake door.
Opposite page top: A view along the port wing root fairing, showing how it blends the fuselage to the wing.
Opposite page bottom: Starboard wing root; note the offset blade antenna behind the cockpit canopy.

FIRE DOOR

Opposite page top: Mid-fuselage detail, starboard side. The NACA-type engine cooling intakes serve also as fire doors and the stencilling nearby is red. Note the offset triangular IFF antenna on the spine, mounted on a flat-topped fairing.
Opposite page bottom: Port upper wing, showing the camouflage pattern and the position of the red/blue roundel.

Above: The port wing from a different angle. The strip of primer just forward of the aileron is probably connected with stress investigation; 'patches' of this type are common on RAF aircraft.
Below: Detail of the 'dog-tooth' wing leading-edge extension, characteristic of all F.6s except the first production machines.

D.899.A.

Opposite page top: The Hunter features a pair of trailing-edge flaps inboard of the ailerons; interior finish is silver. Of interest is the fact that when this aircraft was photographed the port flap was fully retracted. The engine starter access door is open; it has a yellow-finished antenna mounted on the outside, identical to that visible further aft.

Opposite page bottom: The under-surface of the port aileron; the actuator access panel is visible at left.

Above: Hunter F Mk 6A XE606, RAF Brawdy, September 1984. Note the simple canvas intake guard.

Below: A striking photograph of one of No 1 TWU's FGA Mk 9s on final approach, July 1984. The flaps are at full droop and show the 'cut-out' trailing edges necessitated by the installation of the large 230gal drop tanks. The air brake cannot be deployed when the undercarriage is down. *Geoff Lee/British Aerospace*

56

56
52
0
57
54
XG154

Opposite page top: XG228 touches down at RAF St Athan after its final flight from Brawdy. The pilot is Sqn Ldr Ken Becker. Note that the parachute brake has just been streamed. *Geoff Lee/ British Aerospace*

Opposite page bottom: Five Hunters from No 1 TWU, July 1984; all carry the Unit's crest on the nose (but only two the markings of No 79 Squadron) and all are Mk 9s. *Geoff Lee/British Aerospace*

Left: Close view of the nose of XE606. The Aden cannon troughs are painted black, probably to disguise the inevitable staining that would result when the weapons are fired.

Below: Rear three-quarter view of the same aircraft. The jetpipe rim is natural metal.

Left: The Hunter's windscreen is formed of two layers of bullet-proof glass, the canopy being moulded plastic and, in the case of the single-seater's sliding hood, lacking a solid frame at the rear. A minor recognition feature distinguishing the F.6A from the FGA.9 was the absence from the former of the upper IFF antenna on the nose.

Below: A No 1 TWU Hunter T Mk 7. Apart from the remodelled forward fuselage, canopy and dorsal spine, this aircraft differs from the Mk 6A in having a prominent faired outlet midway along the fuselage on each side, a feature associated with 'tropicalised' two-seaters.

Right: Looking into the starboard main intake of an F.6A: the two main features visible here are the compressor face assembly and the central fairing which encloses the liquid-fuel starter. The F.6 introduced the 200-Series Rolls-Royce Avon to the Hunter; offering a thrust of some 10,000lb, it represented a big improvement over the earlier Avon 100s.
Below: Pylon ejector gun fairing and whip antenna, starboard wing, F.6A.
Below right: Detail of port wing-tip navigation light and pitot tube mounting.
Bottom left: Crutching point over starboard inner pylon.
Bottom right: Detail of 'dog-tooth' and pylon brace.

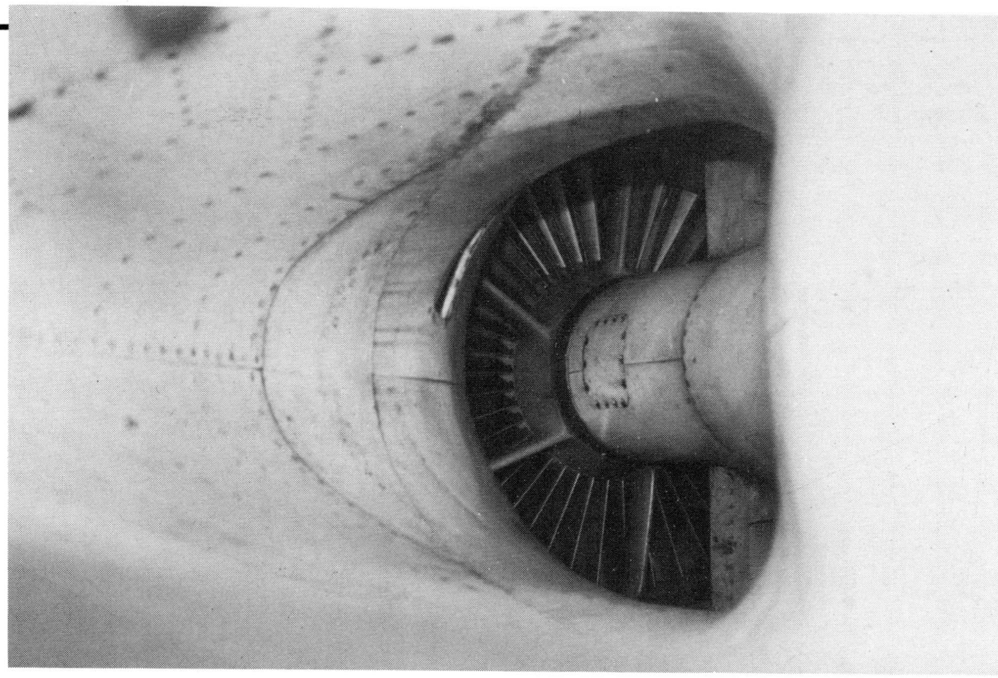

UNDERCARRIAGE

Below: Nosewheel leg and rear nosebay door. General finish is glossy Light Aircraft Grey, with wheel hub silver.
Opposite page top left: Nosegear and rear door from another angle. Note the red-painted ground locking device, secured by butterfly nuts, at the top of the leg.
Opposite page top right: Forward nosebay door, with XE606's individual code clearly displayed.
Opposite page bottom left: Rear view of the front nosebay door, showing the recess to accommodate the wheel.
Opposite page bottom right: Two views of the nosewheel well, the upper photograph showing the front of the bay (note door at far left) and the lower the rear (affording a second look at the ground locking clamp). The general interior finish is silver, with cables, pipes etc in silver, black or red.

INSTALLATION

OXYGEN
CHARGING

MATERIAL
MILD STEEL

GENERAL
ACCESS

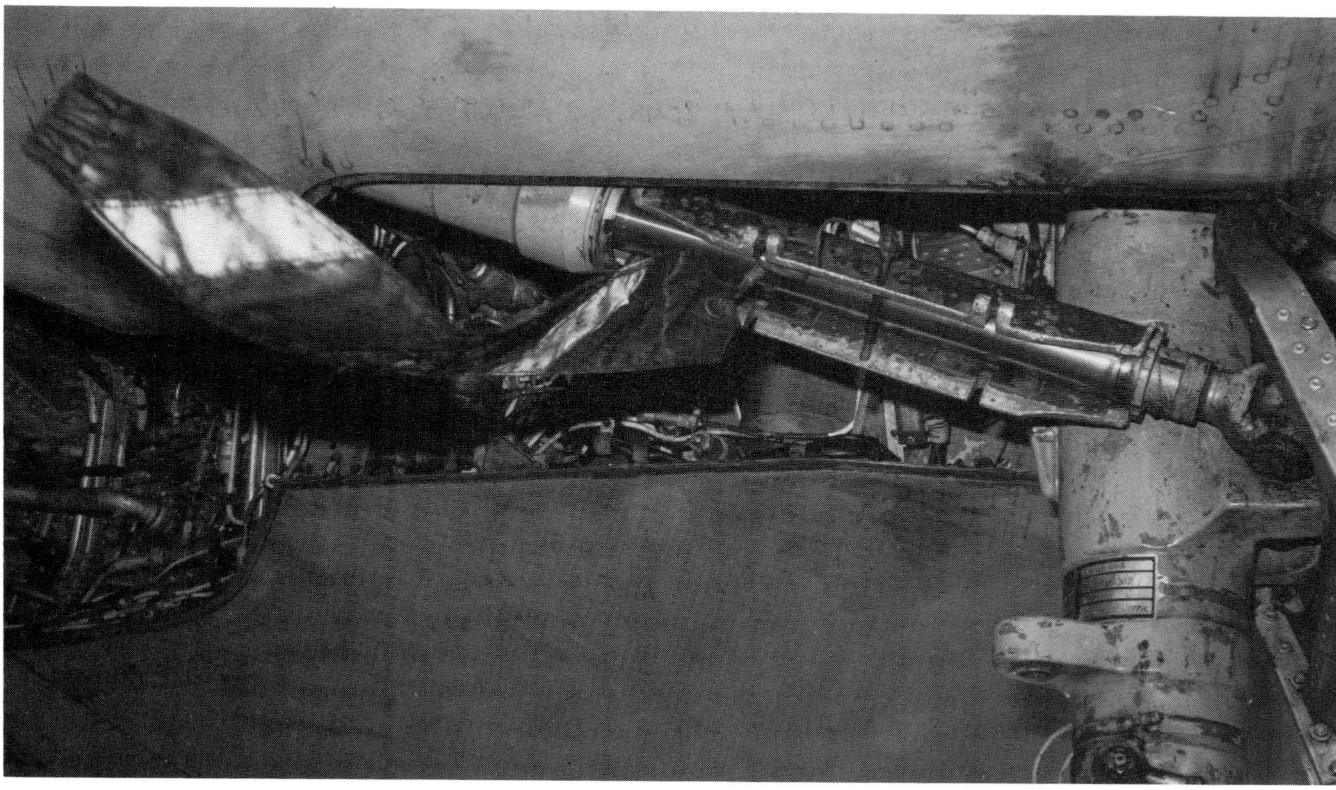

Opposite page top: Two photographs depicting the Hunter's main undercarriage. Note the arrangement of the three-tier outer doors: the lower unit is attached directly to the leg whereas the upper two retract in sympathy with the leg by means of spacer struts. Note also the bright metal oleo section between the torque links.

Opposite page bottom: Inboard main undercarriage door, starboard side. Interior finish is silver.
Above and below: Two views of the starboard main undercarriage bay; note the red and white 'Remove Before Flight' ribbon attached to the main retraction jack. Again, general interior finish is silver.

COCKPIT

Below: A close view of the single-seat Hunter's cockpit canopy reveals the pilot's rear-view mirror mounted at the top of the windscreen frame and, below, the reflector gun sight. A small detail not often appreciated is the row of louvres at the base of the dorsal fairing, immediately behind the canopy.
Bottom: The starboard side console of a Hunter F Mk 6 cockpit. The principal instruments visible here are concerned with the fuel supply and the oxygen system. General interior finish is black. *British Aerospace*
Opposite page: Two views of the main instrument panel, Hunter F Mk 6; the arrangement is very conventional compared to that of more modern military aircraft. Note the gun firing trigger on the front of the control column handgrip. *British Aerospace*
Overleaf: Two photographs depicting the Martin-Baker Type 2H ejection seat fitted to single-seat Hunters; many aircraft were for a time equipped with Type 3Hs during their RAF service. The T Mk 7 ejection seat (Type 4H) was similar in general appearance to that installed in the BAC Lightning (see AEROGUIDE 8). *Martin-Baker Aircraft Co*

CLEAN A/C
OUT BD STORES
IN BD STORES
FUEL
RADAR PRESENTATION
POWER ON ELEVATOR
TRIM
POWER ON AILERON
AILERON
MACH
KNOTS
ALT
FLAPS
FLAPS EMERGY LOWERING
INVERTOR CHANGEOVER
AUTO-LEVELLING ON TAKE-OFF
UP
K.D.G.
OFF
TURN & SLIP SUPPLY NORMAL/EMERGY.
MAXARET FITTED

MASTER
ON
OFF
NAUT. MILES
UP
K.D.G.
RPM
EXH °C
OFF
TURN & SLIP SUPPLY NORMAL/EMERGY.
CAMERA
PORT
FUEL SYSTEM CONTENTS
STBD
FUEL
OIL P.S.I.

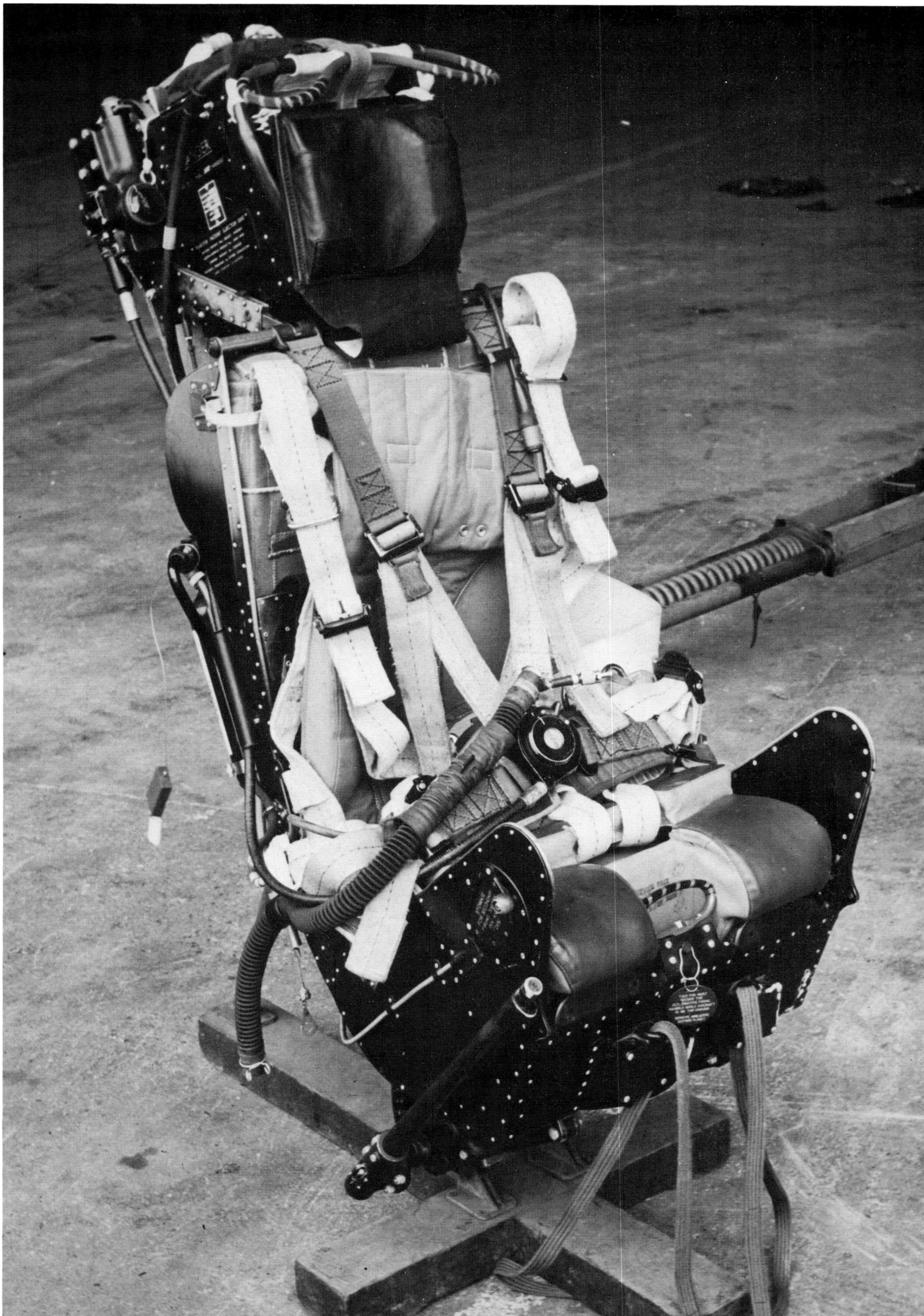

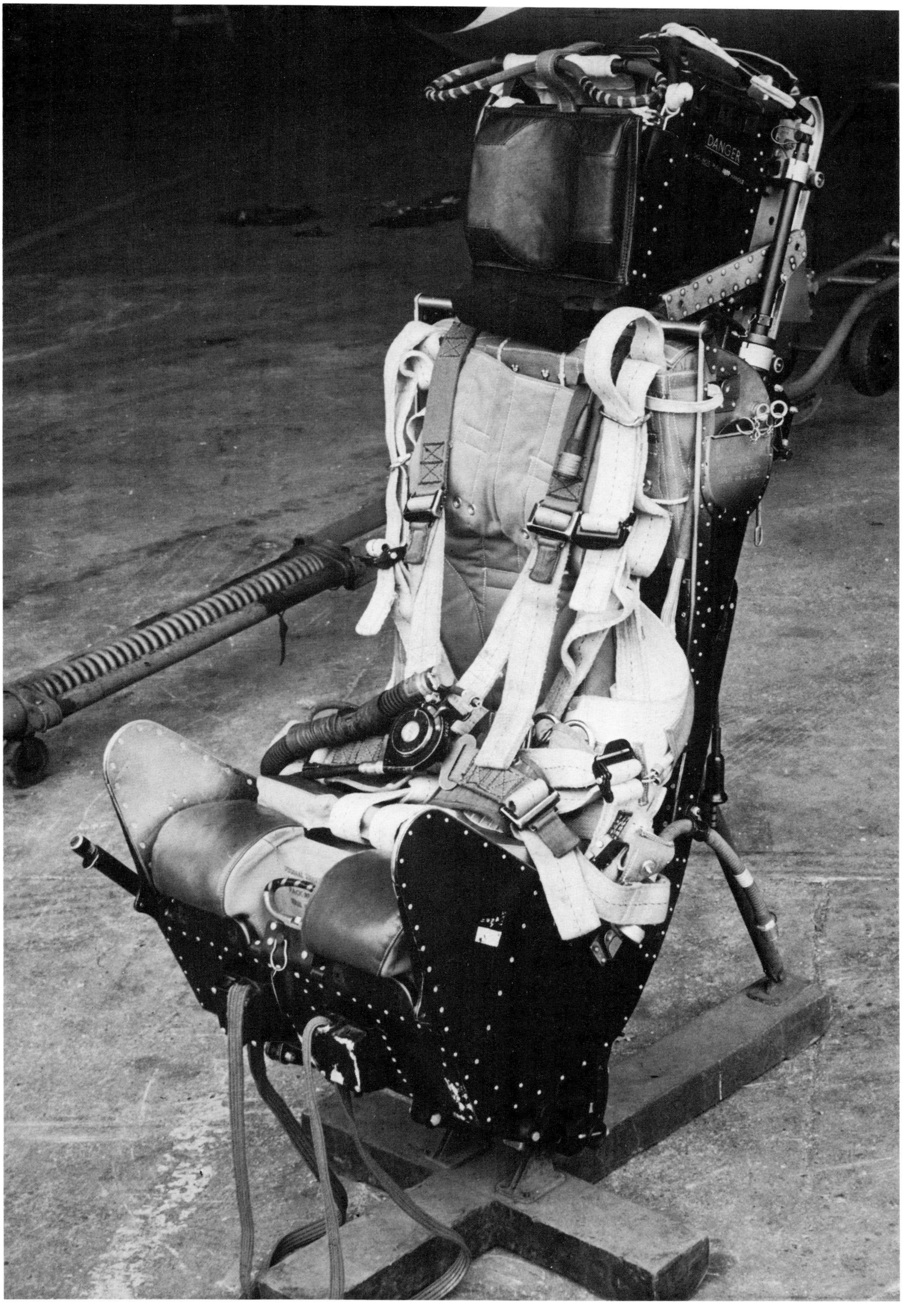
DANGER

WEAPONS & STORES

Below: The Hunter F Mk 6 was cleared to operate a large variety of external stores in addition to its in-built armament of four 30mm Aden cannon (housed in a removable pack – centre of display). Flanking the 30mm ammunition are a pair of 230gal tanks and four 100gal tanks; further forward are 500lb and 1000lb bombs and a pair of practice bomb dispensers. Nearer the aircraft are various rockets – 5in, 3in and pod-mounted FFARs. *British Aerospace*

Bottom: Close-in view of Aden gun troughs, starboard side. Nearby stencilling and sling/trestle marks are black.

Opposite page top: The Hunter's removable Aden cannon pack ensured rapid re-arming between sorties; it contained a maximum of 150 rounds per gun, though fewer were normally carried. *British Aerospace*

Opposite page bottom: Port-side ammunition link collector, with case ejection chutes at rear. These fairings were often referred to as 'Sabrinas', for reasons which will be obvious to those with memories of the 1950s!

1000 LB
SAFE LOAD

HOOD JETTISON
BREAK WINDOW
PULL TO RELEASE
GUN PACK LOCKED
SUPPORT REAR
FUSELAGE BEFORE
REMOVING GUN
PACK
SLING &
TRESTLE
HERE
AIR BOTTLE
CHARGING
LOCKED

Right: The production Hunter T Mk 7 was fitted with only one Aden cannon, and there was thus just a single collector blister, plus this special blast deflector, on the starboard side.
Below: Hunter F Mk 6 trials aircraft fitted with 230gal (inboard) and 100gal drop tanks. *British Aerospace*
Bottom: Two views showing the 230gal tank pylon, 'trough' and sway brace.
Opposite page top: Twenty-four 3in rockets and a 100gal tank seen beneath the port wing of a Hunter F Mk 4. *British Aerospace*
Opposite page centre: Hawker P.1109 (modified F.6) equipped for trials with the DH Firestreak missile. *British Aerospace*
Opposite page bottom: One of the experimental wing-tip tanks fitted to F.6 XG131. *British Aerospace*

SCALE COLOUR PLANS

HAWKER HUNTER F Mk 6A, No 1 TACTICAL WEAPONS UNIT, RAF BRAWDY, SEPTEMBER 1984

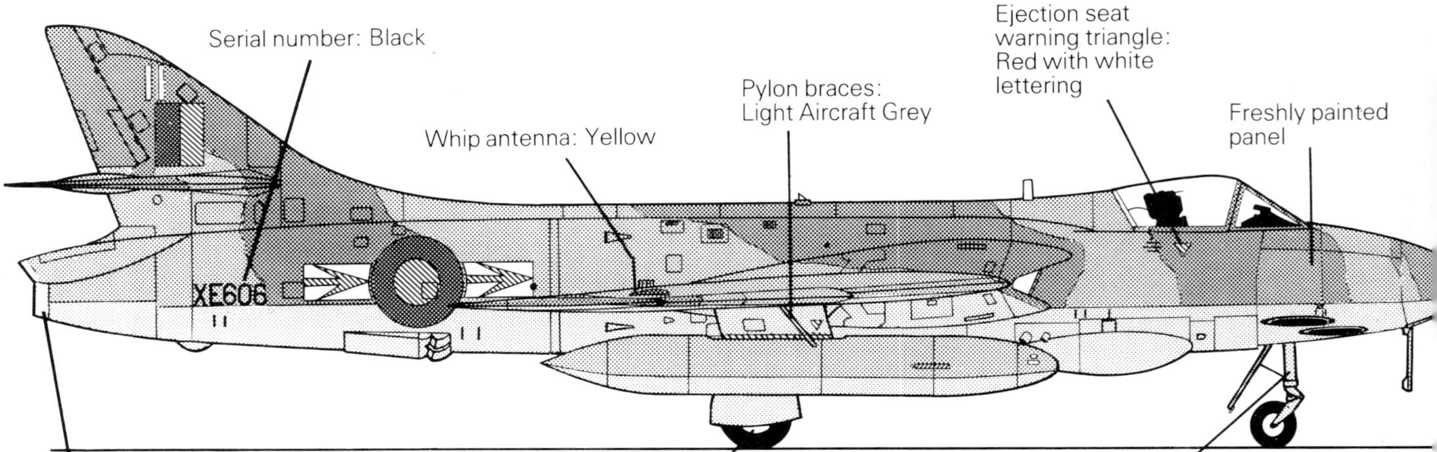

For further details concerning Hunter camouflage and markings, reference to Modeldecal sets 24, 25, 26 and 27 is recommended.

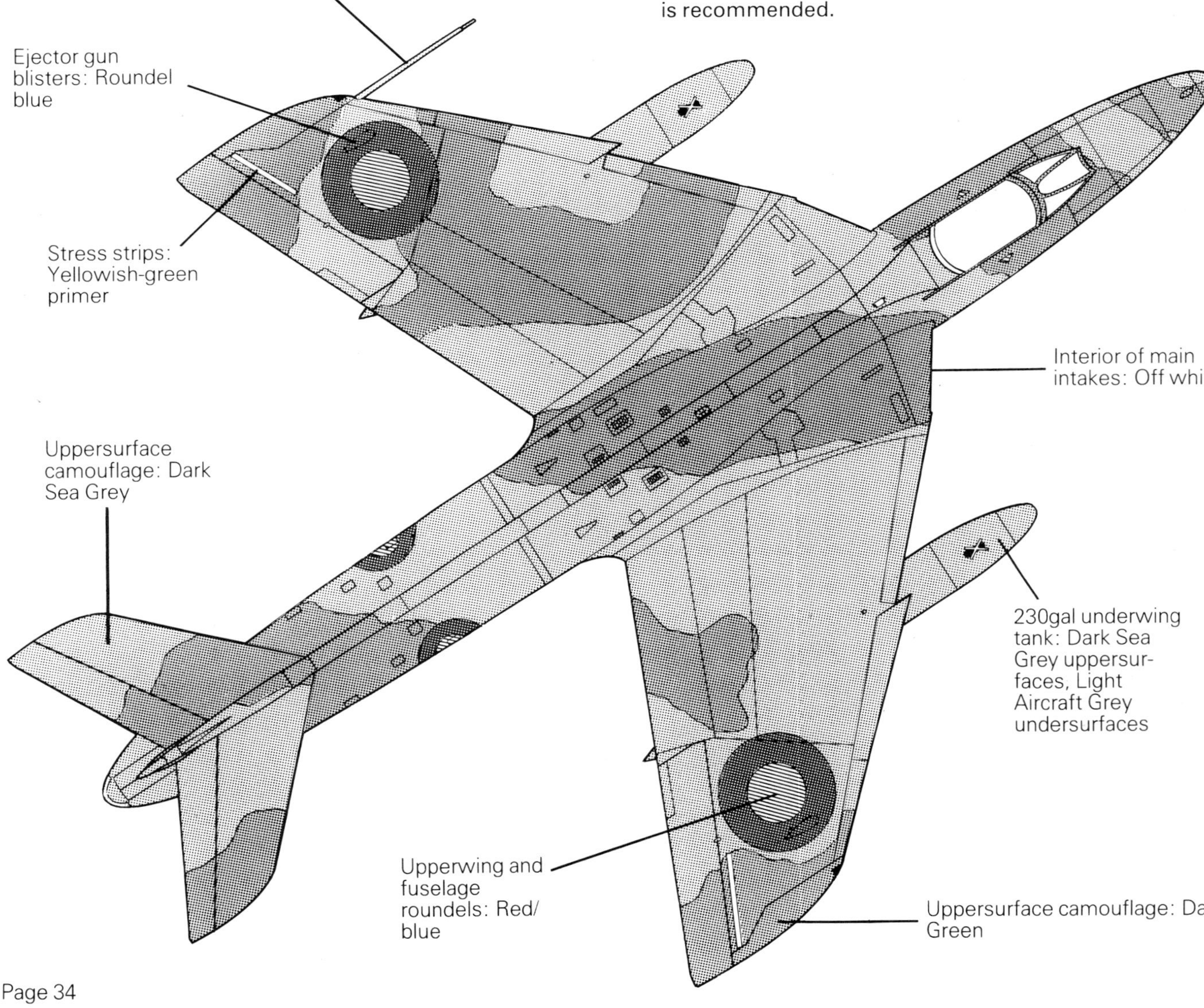

Note that by September 1984 this particular aircraft was showing considerable fading and weathering, especially of the uppersurface paintwork. Many panels appeared to be either replacements or to have been comparatively recently sprayed, resulting in wide variations of hue – see colour section.

1:72 scale

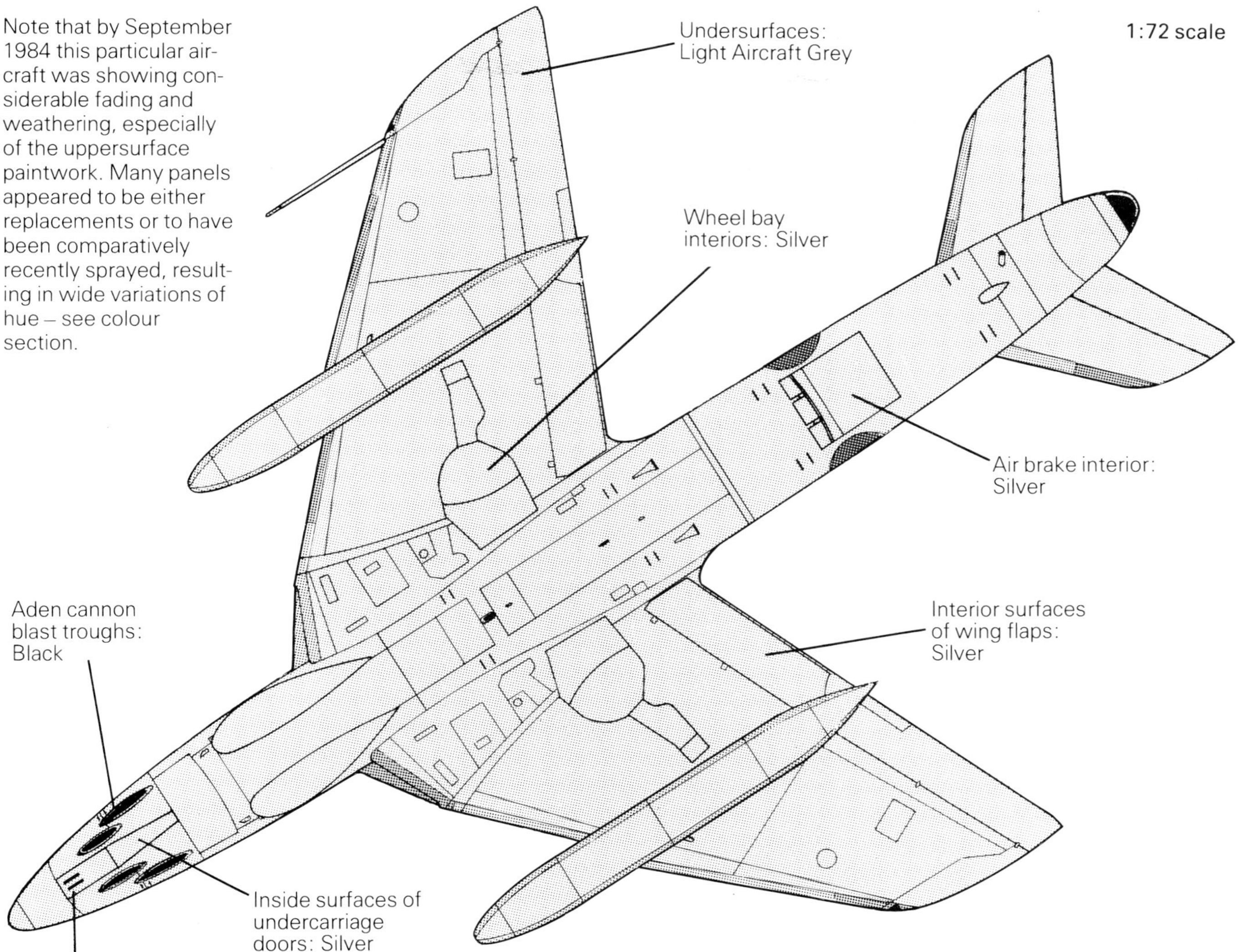

Note that camouflage 'wraps around' wing and tailplane leading edges

Note absence of underwing roundels and serials

British Standard Colour (BSC) references
Dark Sea Grey: BS381C-638
Dark Green: BS381C-641
Light Aircraft Grey: BS381C-627

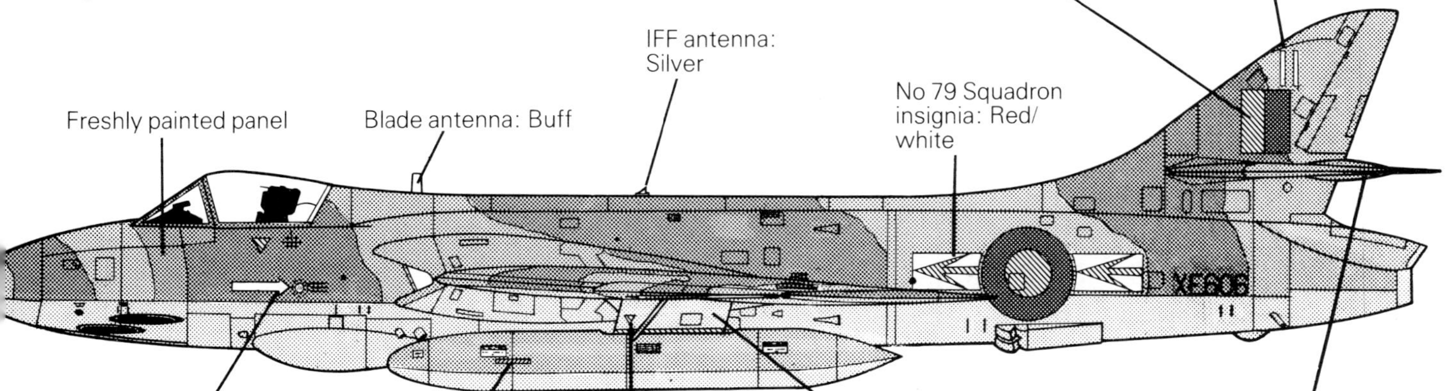

Below: Hunter F Mk 6, No 65 Squadron, RAF Duxford, circa 1960
Dark Sea Grey/Dark Green uppersurfaces, silver undersurfaces; red/white/blue roundels and fin flash; black fuselage and underwing serials; yellow fin code. Nose markings comprise white bars with red chevrons flanking squadron crest.

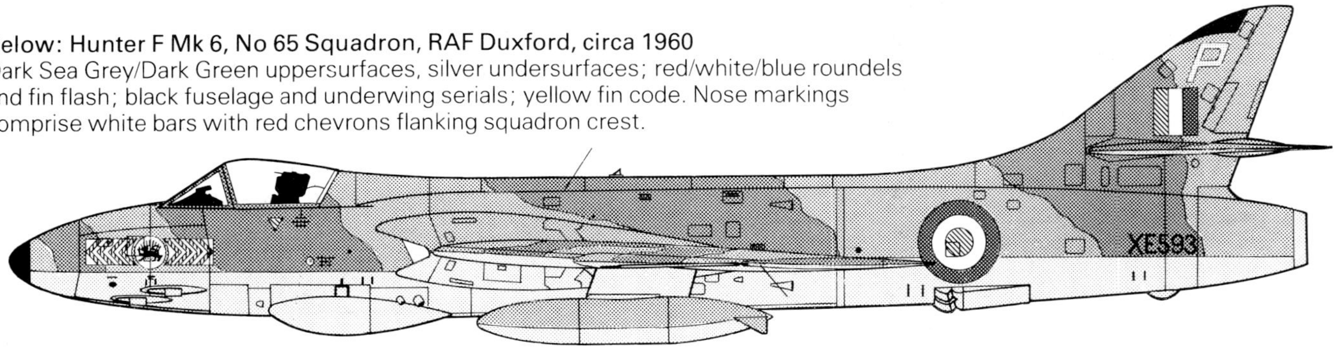

Below: Hunter T Mk 7, No 19 Squadron, RAF Leconfield, circa 1960
Silver overall with yellow trainer bands on fuselage and wings; red/white/blue roundels and fin flash; black fuselage and underwing serials; blue and white nose checks flanking squadron crest.

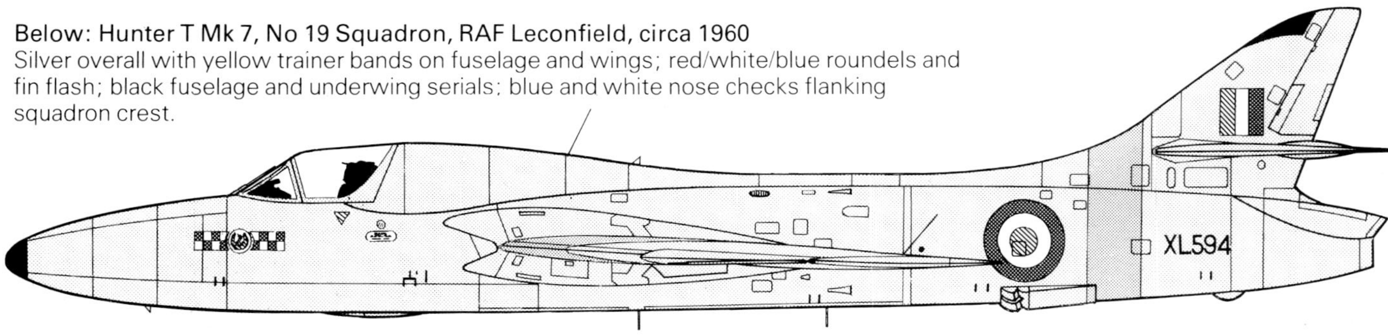

Below: Hunter F Mk 6, No 20 Squadron, RAF Gütersloh, circa 1960
Dark Sea Grey/Dark Green uppersurfaces, silver undersurfaces; red/white/blue roundels and fin flash; black fuselage and underwing serials; white fin code. Squadron markings blue with red (top), white and green stripes.

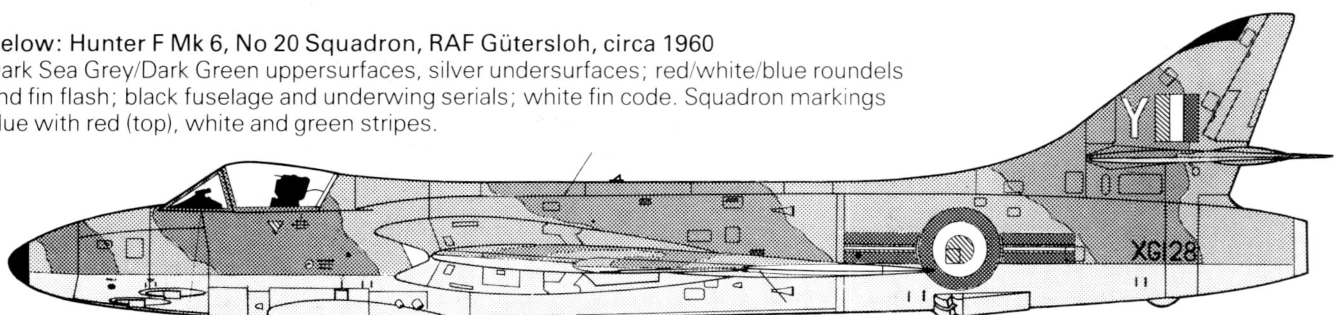

Below: Hunter F Mk 6, No 4 Flying Training School, RAF Valley, 1975
Red/white/Light Aircraft Grey trainer finish; red/white/blue roundels and fin flash; black serials and nose code; 4 FTS emblem on fin.

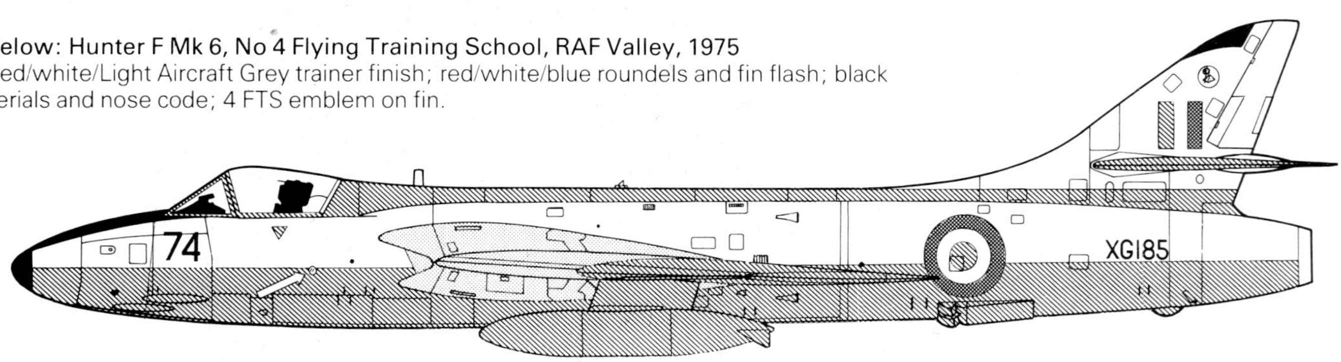

Below: Hunter T Mk 7, No 1 Tactical Weapons Unit, RAF Brawdy, September 1984
Dark Sea Grey/Dark Green uppersurfaces, Light Aircraft Grey undersurfaces; red/blue roundels and fin flash; black fuselage and underwing serials; white tail code; 1 TWU crest on nose.

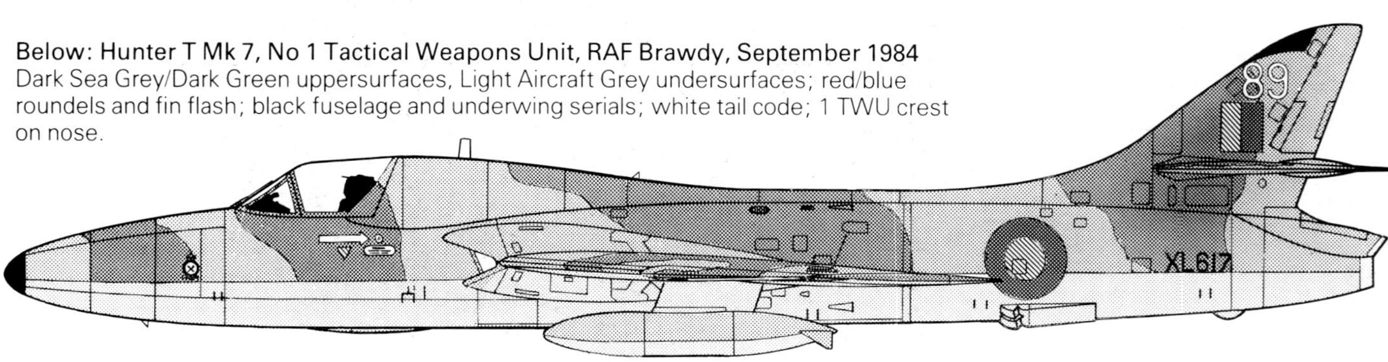